Module 3 Exchange and transport

Topic 1 Exchange surfaces

The need for specialised exchange surfaces

All living organisms need a source of energy: many obtain this energy from the process of respiration. The most efficient form of respiration is aerobic respiration, for which the cells need a supply of oxygen and a means of removing carbon dioxide. With greater activity levels, more oxygen is used up and more carbon dioxide is produced.

In order to provide enough oxygen and remove enough carbon dioxide to satisfy the needs of the organism, a specialised gas exchange site is needed. This must be able to exchange gases quickly enough to provide for the activity of the cells inside the organism. In order to do this, the surface will have certain features, including increased surface area, a thin gas exchange layer and a good blood supply/ventilation to maintain a concentration gradient.

1 Aerobic respiration is one form of respiration, but there is another type that releases energy without the need for oxygen. What is this and why is it useful to organisms such as mammals? (AO1, AO2)   **2 marks**

....
....
....

2 Small organisms rely on diffusion for gas exchange, but larger ones cannot rely on this method of exchange. Explain why diffusion can satisfy the needs of small organisms but it is not suitable for larger organisms. (AO1, AO2)   **2 marks**

....
....
....

3 Explain why mechanisms such as breathing are necessary to keep the gas exchange site refreshed. (AO1, AO2)   **2 marks**

....
....
....

4 Complete the table to explain why a specialised gas exchange surface is needed. (AO1, AO2)

6 marks

Feature	Problem	Adaptation
Concentration gradient	Gases must diffuse down a concentration gradient	
Permeability of surface		
Moist surface		
Surface area to volume ratio	Larger organisms have a reduced surface area to volume ratio and so less gas can diffuse across the surface	

Gas exchange and ventilation in mammals

Breathing movements are needed to ensure an efficient flow of air into the alveoli of the lungs. These movements are a result of the action of the diaphragm and intercostal muscles between the ribs, as well as the natural recoil of the lung tissue itself. There are two main movements involved: inspiration (inhaling) and expiration (exhaling).

A spirometer is a mechanical or electrical device that measures breathing rate, oxygen uptake, tidal volume and vital capacity. The spirometer trace is recorded on a kymograph or on a screen. The trace can be used to determine all the measurements needed when measuring breathing.

1. **Complete the diagram to show the sequence of events involved in one breath, including both inspiration and expiration. (AO1, AO2)** — 8 marks

   | External intercostal muscles _____, raising ribs. Diaphragm contracts and moves _____, pushing organs down |

   ↓

   | Volume of the chest cavity _____, so pressure _____ below atmospheric pressure |

   ↓

   | The lungs _____ as there is now _____ pressure on them |

   ↓

   | Air is _____ the airway |

   ↓

   | External intercostal muscles _____, lowering ribs. Diaphragm _____ and domes up |

   ↓

   | Volume of the chest cavity _____, so pressure _____ above atmospheric pressure |

   ↓

   | The lungs _____ as there is now _____ pressure on them |

   ↓

   | Air is _____ the airway |

2. **Which regulatory mechanism controls inspiration and expiration? (AO1, AO2)** — 2 marks

   ....

   ....

3. **The diagram shows the human breathing system. Name the features labelled A–J. (AO1)** — 5 marks

   A ....

   B ....

   C ....

   D ....

   E ....

   F ....

   G ....

   H ....

   I ....

   J ....

MODULE 3 TOPIC 1 Gas exchange and ventilation in mammals

4 Air sacs are found at the ends of the bronchioles. How is the air sac structure enlarged to provide the large surface area required for a gas exchange site? (AO1, AO2) **2 marks**

....

....

5 Complete the table to summarise the adaptations that are essential in an efficient gas exchange site. (AO1, AO2) **6 marks**

Feature	Requirement	Adaptation
Surface area	Large surface area	
Thickness of the surface	Small diffusion pathway	
Blood supply		
Ventilation mechanism		

6 A number of different tissues make up the structures that form part of the breathing system in mammals. Each one has a specific purpose. For each tissue listed in the table, name the structure in which the tissue is found and state the tissue's specific function. (AO1, AO2) **8 marks**

Tissue	Structure found in	Function
Cartilage	C-shaped rings in trachea; complete rings in bronchi and larger bronchioles of airway	

Tissue	Structure found in	Function
Ciliated epithelium	The walls of the trachea, bronchi and larger bronchioles	
Goblet cells		
Smooth muscle	The walls of the trachea, bronchi and larger bronchioles	
Elastic fibres	Walls of the trachea, bronchi and all bronchioles and alveoli	
Squamous epithelial cells		

 7 If atmospheric air contains approximately 20% oxygen and a subject weighing 60 kg inhaled a volume of air of 1.5 litres in a single breath, how much oxygen would be provided if the subject continued to breathe the same volume of air for the next minute when 20 breaths were recorded for that time period? Calculate the volume of oxygen per kg of body weight. (AO1, AO2)

8 Changes in oxygen levels in the atmosphere have little effect on breathing rate until the level of oxygen falls below approximately 14%, whereas changes in carbon dioxide levels need to rise only a little above the normal level of 0.4% to cause a change in breathing rate. What change would you expect to see if the carbon dioxide level rose above 1%? Give an explanation for your answer. (AO1, AO2) **2 marks**

9 Explain the difference between tidal volume and vital capacity. (AO1) **1 mark**

10 Explain why total lung capacity cannot be measured easily. (AO1, AO2) **2 marks**

11 On a spirometer trace, the vital capacity was read as 4 litres and the tidal volume as 0.5 litres for the first 90 seconds and 1.2 litres for the remaining 90 seconds.

  a When using the spirometer, the subject breathed 19 breaths for the first minute, 21 breaths for the second minute and 24 for the last minute. Calculate the mean tidal volume over the 3-minute period and the mean breathing rate per minute. (AO2, AO3) **1 mark**

  b Suggest the most likely reason for the change in breathing rate and tidal volume over the 3-minute period. (AO2, AO3) **2 marks**

Gas exchange and ventilation in bony fish and insects

The gas exchange systems in bony fish and in insects need the same basic adaptations as are found in mammals. However, these adaptations may be achieved in slightly different ways, depending on the environment. In bony fish, the gills are the gas exchange site. These are subdivided into gill filaments and gill lamellae, which increase the surface area. They are well supplied with blood vessels and are very thin, giving a short diffusion pathway for the oxygen to move from the water to the blood. They are kept moist because the fish live in water. Insects do not transport oxygen in blood. They have an air-filled tracheal system that supplies air directly to all the respiring tissues.

1. Describe in detail the mechanism that keeps the gills supplied with a constant supply of fresh oxygen. (AO1, AO2) **3 marks**

2. Explain how the ventilation mechanism in fish maintains the oxygen concentration gradient and how this assists in gas exchange. (AO2) **2 marks**

Exam-style questions

1. Below are measurements of the surface area and volume of four single-celled organisms.

	Single-celled organisms			
	A	B	C	D
Radius/μm	1	10	100	1000
Surface area/μm²	12.6	1256.7	1.26×10^5	1.26×10^7
Volume/μm³	4.2	4188.8	4.2×10^6	4.2×10^9

a Which organism has the lowest surface area to volume ratio? **1 mark**

b What does this mean in terms of gas exchange for this organism? **2 marks**

2. The gas exchange mechanism in insects is different from that of other animals because it consists of a network of branching tracheae that run through the insect's body. The tubes branch into smaller tubes called tracheoles that terminate in fluid-filled ends.

a What are the main features of an insect's gas exchange site that are the same as those found in other animals? **1 mark**

b Explain the main difference in the gas exchange mechanism of insects when compared with other animals. **2 marks**

c Give *one* advantage of the mechanism found in insects and *one* disadvantage. **2 marks**

Topic 2 Transport in animals

Transport systems in multicellular animals

Transport systems, specialised structures and specialised organs that allow an efficient transport system all become essential once an organism becomes larger, in order to transport essential oxygen, nutrients and waste around the organism. In animals, the transport system is the circulatory system.

1 What are the main reasons that make a transport system necessary? (AO2) *5 marks*

Circulatory systems

In a single circulatory system, the blood flows through the heart once every time it goes around the body, whereas in a double circulatory system the blood flows through the heart twice: once through the right-hand side of the heart in the pulmonary circuit and once through the left-hand side of the heart in the systemic circuit.

Insects have an open circulatory system whereby a single main blood vessel carries blood and delivers it to a body space called a haemocoel. This is unlike a closed circulatory system where the blood flows in blood vessels.

1 The diagram shows the double circulatory system in mammals. Name the features labelled A–H. (AO1) *4 marks*

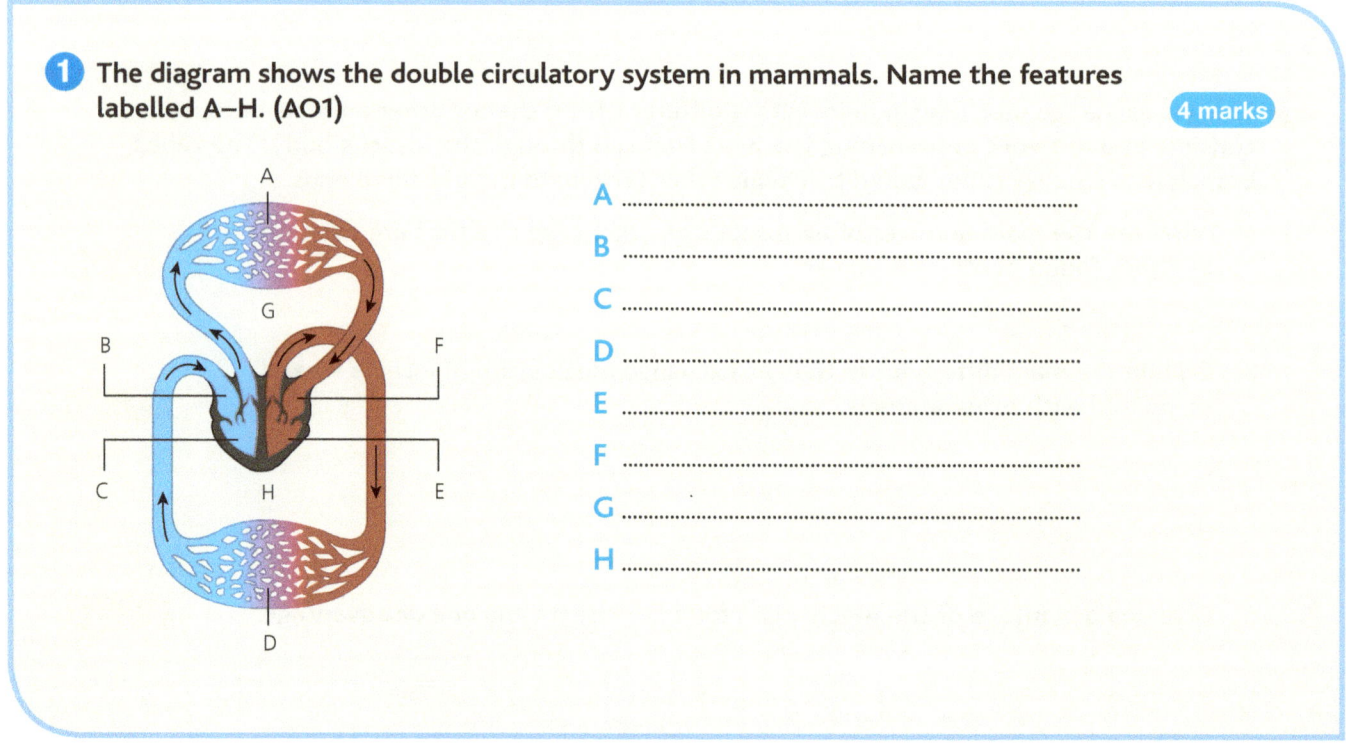

A ....
B ....
C ....
D ....
E ....
F ....
G ....
H ....

2 What advantage is there in having a double circulation rather than a single circulation? (AO1, AO2)
3 marks

....

....

....

3 Fish have a single circulatory system and live in water. Suggest why fish can survive and work efficiently with only a single circulatory system. (AO1, AO2)
2 marks

....

....

....

4 Suggest why the open circulatory system can work efficiently in an insect but would not do so in an animal such as a fish or a mammal. (AO1, AO2)
2 marks

....

....

....

Blood vessels

Blood flows through a series of vessels, each of which is adapted to its particular role in relation to its distance from the heart. The contraction of the heart is vital to ensure a one-way mass flow through the blood vessels to and from the various organs of the body.

1 Complete the table to illustrate the different blood vessels and their role in transport. (AO1, AO2)
4 marks

Type of vessel	Structure	Role in transport
Arteries	Thick wall of collagen (tunica adventitia), thick muscle and elastic tissue (tunica media) and narrow lumen lined with endothelium	
Arterioles		

Type of vessel	Structure	Role in transport
Capillaries	Large number of very small vessels, wall of single squamous cells and a narrow lumen	
Venules		
Veins		

Blood, tissue fluid and lymph

When fluid leaves the blood to form tissue fluid around the cells, almost 80% is returned to the blood in the capillaries. The rest forms lymph, which moves around the body in the lymphatic system before eventually returning to the blood at the neck.

1 Complete the table to compare the structure, components and function of blood, tissue fluid and lymph. (AO1, AO2)

5 marks

Feature	Blood	Tissue fluid	Lymph
Cells			
Fluid			
Dissolved solutes			

Feature	Blood	Tissue fluid	Lymph
Proteins			
Location			

2 The lymphatic system has swellings called lymph nodes. What is their function? (AO1, AO2)

1 mark

....

....

3 Label the diagram to explain how tissue fluid is formed. (AO1, AO2)

5 marks

The mammalian heart and the cardiac cycle

The heart is a muscular pump that is divided into two sides. It pumps blood around the body by regular contractions and is supplied with oxygen and nutrients by the coronary artery, which branches across the heart's surface. The cycle of the heart contracting and relaxing in a regular pattern is called the cardiac cycle.

1. The diagram shows the internal structure of the heart. Name the features labelled A–N. (AO1, AO2) **7 marks**

A ....
B ....
C ....
D ....
E ....
F ....
G ....
H ....
I ....
J ....
K ....
L ....
M ....
N ....

2. What is the significance of the difference in thickness between the following structures?

   a the atria and the ventricles (AO1, AO2) **2 marks**

   ....
   ....
   ....
   ....
   ....

   b the right and the left ventricles (AO1, AO2) **2 marks**

   ....
   ....
   ....
   ....
   ....

3. Name the heart valves and state their function. (AO1, AO2) **3 marks**

   ....
   ....
   ....
   ....
   ....

4 Complete the diagram to show the sequence of events during the cardiac cycle. (AO1, AO2) **7 marks**

| During diastole, the heart chambers are relaxed and pressure is low |

↓

| The atria _____, causing pressure to _____ |

↓

| Blood flows into the _____ and the atria now _____ |

↓

| The ventricles _____ and pressure _____, causing the atrioventricular valves to be forced _____ as the pressure _____ above the pressure in the atria |

↓

| Pressure in the ventricles continues to _____ and puts more _____ on the blood, so the atrioventricular valve _____ slightly into the atrium |

↓

| The semilunar valves _____ as the pressure in the ventricles _____ above that in the arteries |

↓

| Blood flows in the arteries, _____ the pressure within the arteries |

↓

| The ventricles begin to _____ and pressure _____. The semilunar valves _____ when pressure in the aorta is _____ than pressure in the ventricles |

↓

| The atrioventricular valves open and blood that has been flowing into the atria now flows into the ventricles and the cycle starts again |

5 If one complete cardiac cycle takes 0.8 seconds, what is the heart rate? (AO2, AO3) **2 marks**

6 a Define the term *elastic recoil*. (AO1, AO2) **1 mark**

b How does elastic recoil benefit blood flow in the arteries? (AO1, AO2) **2 marks**

7 The heart muscle contracts without any nervous stimulation at all but at its own intrinsic beat. Name this type of intrinsic beat. (AO1) — 1 mark

8 In order to control the heart rate, the heart contraction is controlled by signals from the autonomic nervous system.

   a Name the two parts of the autonomic nervous system. (AO1, AO2) — 1 mark

   b Explain the role that each part plays in coordination of the heartbeat. (AO1, AO2) — 3 marks

9 Complete the table to illustrate the electrical coordination of the heart beat. (AO1, AO2) — 5 marks

Name of structure	Position	Role
	Pacemaker set in the wall of the right atrium	
Non-conductive septum	Between the atria and ventricles	
	Conductive node at the base of the atria in the septum	
	Specialised muscle fibres in the central septum of the heart	
Purkyne tissue		

10 The graph shows an ECG of one beat of a healthy heart.

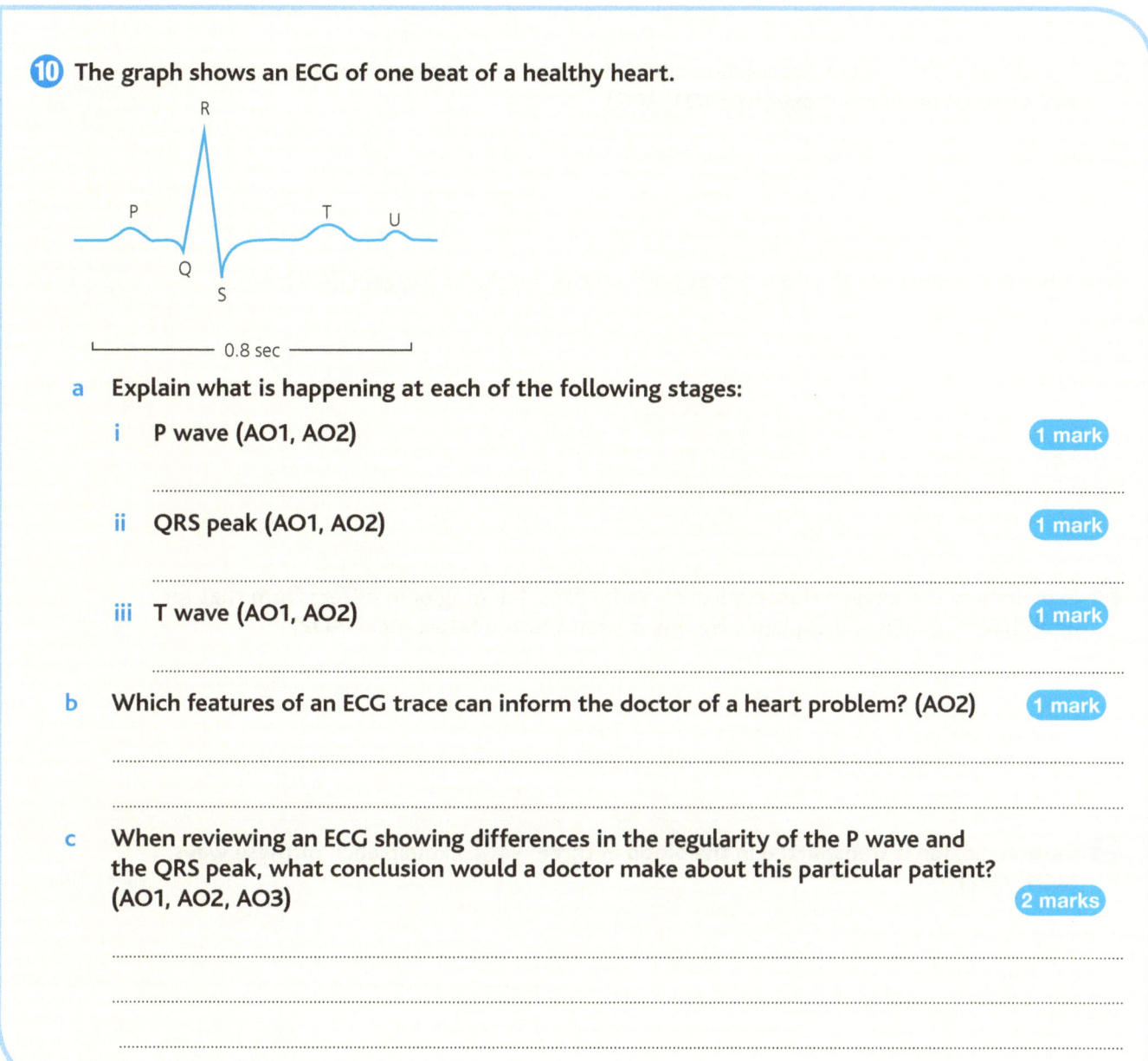

a   Explain what is happening at each of the following stages:

   i   P wave (AO1, AO2) — 1 mark

   ii QRS peak (AO1, AO2) — 1 mark

   iii T wave (AO1, AO2) — 1 mark

b   Which features of an ECG trace can inform the doctor of a heart problem? (AO2) — 1 mark

c   When reviewing an ECG showing differences in the regularity of the P wave and the QRS peak, what conclusion would a doctor make about this particular patient? (AO1, AO2, AO3) — 2 marks

Haemoglobin

Haemoglobin transports oxygen round the body in the red blood cells, as well as transporting carbon dioxide back to the lungs from the respiring cells. Oxygen transport relies on the increased affinity of haemoglobin for oxygen as the partial pressure of oxygen increases.

In the lungs, where there is plenty of oxygen (a high partial pressure of oxygen), haemoglobin takes up oxygen easily and gives it up less easily. In the respiring tissues, where the partial pressure of oxygen is low, oxygen is released readily. This is best shown in the oxygen dissociation curve.

1 a Describe the structure of the haemoglobin molecule, giving details of its protein structure and prosthetic group. (AO1, AO2) — 2 marks

  b State the importance of this molecule. (AO1, AO2) — 2 marks

2 Explain why the oxygen dissociation curve shows a steep increase between 2 and 4 kPa partial pressure of oxygen. (AO1, AO2) `3 marks`

....
....

3 What influence does the Bohr effect have on the uptake of oxygen? (AO1, AO2) `3 marks`

....
....
....
....
....

4 Explain how the oxygen dissociation curve for fetal haemoglobin differs from that for adult haemoglobin and explain why this is useful to the fetus. (AO1, AO2) `2 marks`

....
....
....

5 Carbon dioxide is transported in the blood in three ways. Explain each of these ways. (AO1, AO2) `3 marks`

....
....
....
....
....
....

Exam-style questions

1 a Tick the box in the table to indicate the most accurate description of the mammalian circulatory system. `1 mark`

	Open circulatory system	Closed circulatory system
Single circulatory system		
Double circulatory system		

b Which terms are used to describe the highest and the lowest blood pressure? `2 marks`

c Blood pressure fluctuates between 18.5 kPa and 13.5 kPa as it flows along the aorta. Explain what causes these fluctuations. `2 marks`

d As the blood flows from the arteries through the capillaries and the veins, the blood pressure falls considerably. How is the blood able to continue flowing in the vessels back to the heart? `2 marks`

e Explain why it is important that blood pressure changes as it flows from the aorta to the capillaries. `2 marks`

2 Complete the passage by filling in the missing words. `5 marks`

Haemoglobin is a pigment found in red blood cells. These cells are also known

as ……………………………… . Haemoglobin has a high ……………………………… for

oxygen. In the lungs, the haemoglobin associates with oxygen to form ……………………………… .

In respiring tissues, the oxygen is released by dissociation. In very active tissues, the amount

of oxygen released can be increased by the presence of more……………………………… .

This is known as the ……………………………… effect.

3 Describe how the contraction of the heart is initiated and coordinated. Use appropriate technical terms spelt correctly in your answer. `5 marks`

Topic 3 Transport in plants

Transport systems in multicellular plants

Plants need a transport system to transport water and ions from the soil via the roots to the aerial parts of the plant. It is also needed to transport nutrients, such as glucose made in the leaves, to the rest of the plant. In multicellular plants there are two separate systems: a water transport system in the xylem and a system to transport assimilates (nutrients) in the phloem.

1 Plants do not need to transport oxygen and carbon dioxide. Explain why not. (AO2) **3 marks**

....
....
....
....

2 Water is transported up through the plant by a tissue called xylem. This consists of a system of tubes called xylem vessel elements. Describe the main features of the xylem vessel elements that demonstrate how they are adapted to transport water. (AO1) **3 marks**

....
....
....
....

3 Name *two* types of xylem cells other than vessel elements. (AO1) **1 mark**

....

4 The lignin thickening in xylem vessels may be in a solid layer or in a pattern of lignin such as rings, spirals or strips. What is the advantage of each type?

a a solid layer (AO2) **1 mark**

....

b a broken layer such as rings or spirals (AO2) **1 mark**

....

The vascular system

Assimilates and minerals are transported both up and down the plant in a tissue called phloem. This consists of a system of tube called sieve tube elements and support cells called companion cells.

Water movement occurs between the cells because there is a difference in water potential. This is also true between cells and the atmosphere and between cells and the soil water. Water moves down a water potential gradient from an area of higher water potential to an area of lower water potential.

1. Describe how the sieve tube elements are adapted to carry out the function of transporting assimilates. (AO1, AO2) — 3 marks

2. Give *three* differences between xylem vessels and sieve tube elements. (AO1) — 3 marks

3. Complete the table to show the three main routes of water transport from the root hair to the xylem. (AO1) — 2 marks

Method of water transport	Details of route	Function
		Most of the water uses this route; it is efficient as water moves rapidly as if through filter paper
	Through the cytoplasm of one cell to another using the plasmodesmata in side walls	
Vacuolar		

4. What is the role of the Casparian strip in the walls of the endodermal cells? (AO2) — 2 marks

Transpiration

Transpiration is the loss of water vapour from the upper parts of a plant, mainly the leaves. Although some water evaporates and diffuses through the leaf surface, most water vapour is lost via the stomata.

1. In the xylem, water and dissolved ions move upwards. There is a small pressure exerted by the roots (root pressure) to assist water movement upwards, but this accounts for only a little movement.

   a. What is the main mechanism that allows water to move up the xylem? (AO1, AO2) — 1 mark

21

b Give details of this mechanism. (AO1, AO2) .... 2 marks

....

....

....

....

2 Which features are required in order to allow the water transport system to work efficiently? (AO1, AO2) .... 2 marks

....

....

....

....

3 Why is transpiration sometimes described as 'a necessary evil'? (AO1, AO2) .... 1 mark

....

4 Complete the table to show the main factors that affect the rate of transpiration out of a leaf and how they affect transpiration rate. (AO1, AO2) .... 3 marks

Factor	Requirement	Effect on transpiration rate
Temperature	Needs a water potential gradient from inside leaf air spaces to outside the leaf; the stomata must be open — high temperatures cause stomatal closure which stops transpiration	
	The water potential gradient must be higher inside leaf than outside to allow water to move down the gradient; the stomata must be open	
	The diffusion gradient between air outside leaf and the air in the leaf air spaces must be high so that water can diffuse out; the stomata must be open	
	The stomata must be open — in the dark they are closed but open as light increases	

5 A potometer is used to measure water uptake.

a What assumption must be made when using a potometer to measure transpiration? (AO3) .... 1 mark

....

....

b Give two reasons why this is not a valid assumption. (AO3) .... 2 marks

....

....

....

....

6 When setting up a potometer of any design, what essential steps must be taken to ensure that reliable results are obtained? (AO3) — 2 marks

7 A potometer was set up with a small leafy shoot and placed in three different temperature conditions. The air movement and humidity around the apparatus were kept constant. The distance moved by the bubble in the capillary tube was measured over 3 minutes at each temperature and the following results were recorded.

Temperature/°C	Distance moved by bubble/mm			
	1 min	2 min	3 min	Mean distance/mm min^{-1}
20	34.0	21.0	30	
30	56.0	57.0	53.0	
40	28.0	27.0	26.0	

a Complete the table by calculating the mean distance moved by the bubble at each temperature. (AO3) — 1 mark

b Use the data in the table to describe the effect of temperature on the distance moved by the bubble. (AO3) — 2 marks

c Comment on the recording of the results at 20 °C. (AO3) — 2 marks

d Suggest a reason for the results obtained at 40 °C. (AO3) — 1 mark

e What method would you use to decide if one of the results were anomalous? (AO3) — 1 mark

8 Another potometer was set up using a capillary tube with a bore diameter of 1.2 mm. The bubble moved by 25.0 mm in 30 seconds. Calculate the water uptake as mm^3 min^{-1} for this shoot. (AO3) — 2 marks

Adaptations of plants to the availability of water

All land-based plants must control water loss to some extent to avoid plasmolysis and cell death. Xerophytes must severely control water loss because they inhabit dry conditions, whereas hydrophytes such as water lilies are adapted to living in water.

1 Describe the mechanisms used by most land-based plants to control water loss. (AO1, AO2) — 2 marks

2 Give *four* xerophytic adaptations that allow plants to survive in conditions that otherwise would be hostile in terms of water loss. (AO1) — 4 marks

3 Give *four* adaptations that a hydrophyte shows to allow it to survive in its habitat. (AO1) — 4 marks

Translocation

Translocation is movement of soluble assimilates that are formed during photosynthesis in the leaves. Sucrose is the main molecule involved, but other molecules such as amino acids and fatty acids are also formed as needed. Movement is from the source, where they are produced, to the sink, where they are needed.

1 The main source for the production of assimilates is the leaves. Give *three* examples of the main sinks in a growing plant. (AO1, AO2) — 3 marks

2 The mass flow hypothesis is used to describe the mechanism for transport in the phloem. Explain why this is called a hypothesis. (AO2, AO3) — 1 mark

3. The mechanism proposes that sucrose is loaded into the phloem at the leaves and unloaded at the sinks.

   a. Describe the mechanism for loading sucrose into the phloem. (AO1, AO2) **2 marks**

   b. Describe how sucrose is unloaded at the sinks. (AO1, AO2) **2 marks**

4. Review the evidence for the mass flow theory and state four points that strongly support the theory. (AO1, AO2) **2 marks**

Exam-style questions

1. A student used a potometer to measure the rate of water uptake of a cut shoot of the sycamore tree, *Acer pseudoplatanus*. The student placed the potometer on a balance to measure the rate of water loss. The following results were recorded.

Time/min	Mass of potometer and shoot/g	Volume of water in graduated tube/cm^3
0	376.8	10.00
4	376.6	9.90
8	376.4	9.75
12	376.2	9.65
16	376.0	9.50
20	375.8	9.40

   a. Which precautions should be taken to ensure that there are no air blocks in the cut end of the sycamore stem? **2 marks**

   b. Suggest the advantages of using potometers that measure both mass loss and water uptake. **2 marks**

c i Use the data in the table to calculate the rate of water loss and the rate of water uptake per hour. Show your working. *2 marks*

ii State the assumption that is made in using mass loss as a measurement of water loss. *1 mark*

iii The student concluded that the shoot was at the point when it would wilt. State the evidence for that conclusion. *1 mark*

iv Suggest why wilting may be advantageous to a plant. *2 marks*

2 The cells in the epidermis of a plant root are specialised to absorb minerals from the surrounding soil. State the process by which root epidermal cells absorb minerals from the soil and describe how these cells are specialised to achieve absorption. *3 marks*

3 Describe the xerophytic features of a marram grass leaf and explain how each feature reduces loss of water vapour. *5 marks*

Module 4 Biodiversity, evolution and disease

Topic 4 Communicable diseases, disease prevention and the immune system

Pathogens

Pathogens are microorganisms that cause disease. There is a wide range of them living in a wide of range of habitats. They transmitted in a variety of ways, both direct and indirect. Pathogens include:

- bacteria such as tuberculosis (TB), bacterial meningitis, ring rot (potatoes, tomatoes)
- viruses such as HIV/AIDS (human), influenza (animals), tobacco mosaic virus (plants)
- protoctista such as malaria, potato/tomato late blight
- fungi such as black sigatoka (bananas), ringworm (cattle), athlete's foot (humans)

1. Describe how the bacteria living on and in the human body are considered to be beneficial? (AO1) *(1 mark)*

2. How is a parasite different from beneficial organisms? (AO2) *(1 mark)*

3. What is the difference between a communicable (infectious) disease and a non-communicable disease? (AO1, AO2) *(2 marks)*

4. Disease transmission is the transfer of a pathogen from one host to another. Give *three* examples of different methods of disease transmission. (AO1) *(3 marks)*

5 Tuberculosis (TB) is a communicable human disease that infects the lung and other tissues. It had been largely eradicated in the world, but is once again infecting humans and its spread is increasing. Suggest why the ability of TB to infect is increasing around the world. (AO2) `2 marks`

....

....

6 *Neisseria meningitides* is one bacterium that causes bacterial meningitis. Describe how this bacterium is different from others in the way that it infects the body. (AO1) `1 mark`

....

7 Bacterial pathogens of plant tissues often kill their hosts. How do these bacteria survive once the host has been killed? (AO1) `1 mark`

....

8 Why are viruses sometimes described as 'ultimate parasites'? (AO1) `1 mark`

....

9 Influenza is one type of virus that causes worldwide epidemics and many deaths. The human immunodeficiency virus (HIV) is another human virus. State two similarities between these two viruses and give one major feature of HIV that is not found in the influenza virus. (AO1, AO2) `3 marks`

....

....

....

....

10 a Give the correct scientific name for the protoctistan that causes malaria. (AO1, AO2) `1 mark`

....

b How is its method of infection different from many other disease organisms? (AO1, AO2) `2 marks`

....

....

11 Although they are eukaryotes, fungi have a number of distinctive features that differentiate them from both plants and animals. What are the characteristics of the fungi that classify them in a different group? (AO1) `2 marks`

....

....

12 There are many more plant fungal diseases than animal fungal diseases. Suggest why this may be the case. (AO1, AO2) *2 marks*

....

....

13 Give *two* examples of plant fungal diseases and *one* example of an animal fungal disease. (AO1) *3 marks*

....

....

....

14 Explain why the fungus that causes potato blight is not considered to be a true fungus. (AO1, AO2) *2 marks*

....

....

....

15 Complete the table to compare the method of transmission for some pathogens. (AO1, AO2) *5 marks*

Pathogen		Disease	Transmission
Bacterial	Animal	TB — *Mycobacterium tuberculosis* and *M. bovis*	
	Plant	Ring rot — *Clavibacter michiganensis* subsp. *sependonicus*	
Viral	Animal	Influenza — influenza A, B or C (from family *Orthomyxoviridae*)	
	Plant	Tobacco mosaic virus — TMV	
		Peach blight and potato leaf roll	

Pathogen		Disease	Transmission
Protoctistan		Malaria — *Plasmodium* species (e.g. *P. falciparum*)	
Fungal	Animal	Ringworm — *Trichophyton verrucosum* or athlete's foot — *Trichophyton rubrum*	
	Plant	Black sigatoka — *Mycosphaerella fijiensis*	
Fungal-like		Potato blight — *Phytophthora infestans*	

16 Explain what is meant by:

   a direct transmission (AO1, AO2) — 1 mark

   ....

   ....

   b indirect transmission (AO1, AO2) — 1 mark

   ....

   ....

17 In a person infected with HIV, the virus eventually begins to attack the host's lymphocytes and reduces their number. This is often diagnosed only when opportunistic diseases begin to develop. What is an opportunistic disease? (AO1) — 1 mark

   ....

   ....

   ....

18 A researcher carried out practical investigations to determine the resistance of a recent strain of *Mycobacterium tuberculosis*. The bacteria were cultured and grown in a suitable culture medium and then transferred to two culture plates, each containing the culture medium and a combination of two of the four main drugs used to treat TB. One plate contained isoniazid and pyrazinamide. The other plate contained rifampicin with ethambutol. The researcher found that drug-resistant TB pathogens grew on both plates.

However, when she repeated the experiment using culture medium containing three of the four drugs, only one small colony of bacteria developed. She concluded that using three drugs to treat TB would prove to be the best method for controlling infection. What other steps should she have taken in order to be sure that her conclusions were valid? (AO2, AO3) **3 marks**

....
....
....
....
....

Factors affecting transmission

Diseases that are endemic are always present in a population. The pathogen is always present in that area and in that population and it will begin to spread quickly whenever the population is weakened by other factors. Disease outbreaks may also occur whenever a new strain of the pathogen occurs.

Some animals and plants are resistant to certain diseases. They have inherited a gene that makes them less susceptible to those diseases, for example individuals with a single gene for sickle-cell disease are resistant to malaria.

1 What is the difference between resistance and immunity to a disease? (AO1, AO2) **2 marks**

....
....
....

2 Complete the table to show the factors affecting the transmission of certain types of disease. (AO1) **3 marks**

Type of transmission	Type of disease	Factors that affect transmission
	Influenza	
	TMV in plants	Monoculture farming methods
Indirect	Transmission by vectors	Climate and weather; a breeding ground such as stagnant water

Type of transmission	Type of disease	Factors that affect transmission
Human diseases	H1N1 influenza strain or SARS	
	Cholera, typhoid or polio	
	Smallpox	
	HIV	

Plant defences

Plants are surrounded by pathogens and have evolved physical and chemical defences against them. Physical defences include cellulose cell walls, lignin thickening of the cell walls, waxy cuticles and bark. Some chemical defences are already present to prevent entry of a pathogen, whereas others are released only after a pathogen has been detected.

1 How do physical defences help the plant to defend itself against infection? (AO1) *1 mark*

2 List *three* chemical defences and in each case state how it prevents infection. (AO1, AO2) *3 marks*

3 Some plants have active defence mechanisms that are stimulated when a pathogen enters. Give *one* example of such a mechanism. (AO1, AO2) *2 marks*

4 What is systemic-acquired resistance? (AO1) *1 mark*

The primary non-specific defences in animals

The transmission of pathogens from one organism to another occurs in a number of ways and the body has a variety of methods to combat this spread. Some of the mechanisms are quite simple, whereas others are very sophisticated.

1 Complete the table to show the four main methods of defence that animals use to defend themselves against disease. (AO1, AO2)   *7 marks*

Category of defence mechanism	Method	Example
Cellular	Cells signal the body about an invasion of pathogens; produce substances that protect; ingest pathogens	
	The body secretes substances that change the environment to become hostile to the pathogen; causes pathogens to burst; stops pathogens reproducing or growing; prevents entry into the cells	The enzyme lysozyme is an antibacterial agent that is produced in tears; fatty acids in sebum have antibacterial properties; hydrochloric acid produced by the stomach lining destroys bacteria; histamines; cytokines
	Tissues act as a barrier	
Microbiome (our own bacterial flora living on and within us)		

2 The first line of defence is effective and involves barrier, chemical and expulsive reflex methods.

a Explain how the skin acts as both a barrier and a chemical defence. (AO1, AO2)   *2 marks*

....

....

....

b What is an expulsive reflex and why is it important? (AO1, AO2)   *2 marks*

....

....

3 Give *one* example of an unintentional effect on the body's defence of taking antibiotics for an infection. (AO1) `1 mark`

4 a List the *five* aspects of the second line of defence present in the body and for each give at least *one* example of how it acts in defence of the body. (AO1) `5 marks`

b Explain why this is called a non-specific defence mechanism. (AO2) `2 marks`

5 What is the third line of defence? (AO1) `1 mark`

The primary and secondary immune responses

The immune response is the sequence of events in response to the presence of a pathogen that results in the selection and production of increased numbers of specialised lymphocytes. It adapts to changes in the environment and to different pathogens.

The primary response is triggered the first time a pathogen invades the body. This takes a few days as the specific B and T lymphocytes must be selected. Therefore, the maximum concentration of antibodies is not achieved until a number of days after infection. On any subsequent invasion by the same pathogen, the secondary response is triggered. This response is much more effective because the blood carries many memory cells that are specific to this pathogen.

1 What are the main differences in origin and function between B lymphocytes and T lymphocytes? (AO1) `4 marks`

....
....
....
....

2 Complete the diagram to show the sequence of events in B lymphocyte activation. (AO1) `5 marks`

[ ]
↓
[ ]
↓
[ ]
↓
[ ]
↓
[ ]

3 Outline the similarities and differences between B lymphocyte activation and T lymphocyte activation. (AO1, AO2) `6 marks`

....
....
....
....
....
....
....
....

4 a Explain why the secondary immune response is different from the primary immune response. (AO1, AO2) `2 marks`

....
....
....

MODULE 4 TOPIC 4 The primary and secondary immune responses

b Explain what the main differences are. (AO1, AO2) — 2 marks

....
....
....

5 The graph shows the change in concentration of antibodies during the primary and secondary immune responses.

a What is the response time (in days) from the time of infection for the primary and secondary responses to reach their maximum level? (AO1, AO2) — 1 mark

....

b Estimate (in arbitrary units) the difference in antibody concentration for both the primary and the secondary infection at the maximum level for each. (AO1, AO2) — 1 mark

....
....

Antibodies

Antibodies are proteins in the plasma called immunoglobulins. They have a quaternary structure and are composed of a constant region and variable regions.

1 Explain why antibodies are said to have a quaternary structure and state what form of bonding is used to form the typical Y shape of the antibody. (AO1) — 2 marks

....
....

2 What is the function of the following?

a the constant region of an antibody (AO1) — 1 mark

....
....

b the variable region of an antibody (AO1) `1 mark`

3 Name the *three* main classes of antibodies and explain how each works to destroy pathogens. (AO1, AO2) `3 marks`

Types of immunity

Natural active immunity occurs when someone is exposed to a live pathogen, such as the result of catching a flu virus from someone who sneezes. Natural passive immunity may be acquired through breast milk or the mother's placenta.

Artificial active immunity can result from being given a vaccine — a substance that contains the antigen. The vaccine stimulates a response against the antigen without causing symptoms of the disease. Artificial passive immunity is a short-term immunisation by the injection of antibodies that are not produced by the recipient's cells.

1 Complete the table to compare natural immunity and artificial immunity. (AO1, AO2) `4 marks`

Type of immunity	Active	Passive
Natural		
	e.g. chicken pox virus	e.g. crossing the placenta during pregnancy and in the mother's milk during breast-feeding
Artificial		
	e.g. mumps or whooping cough	e.g. tetanus, rabies or diphtheria

2. The graph shows the change in antibody concentration in passive immunity.

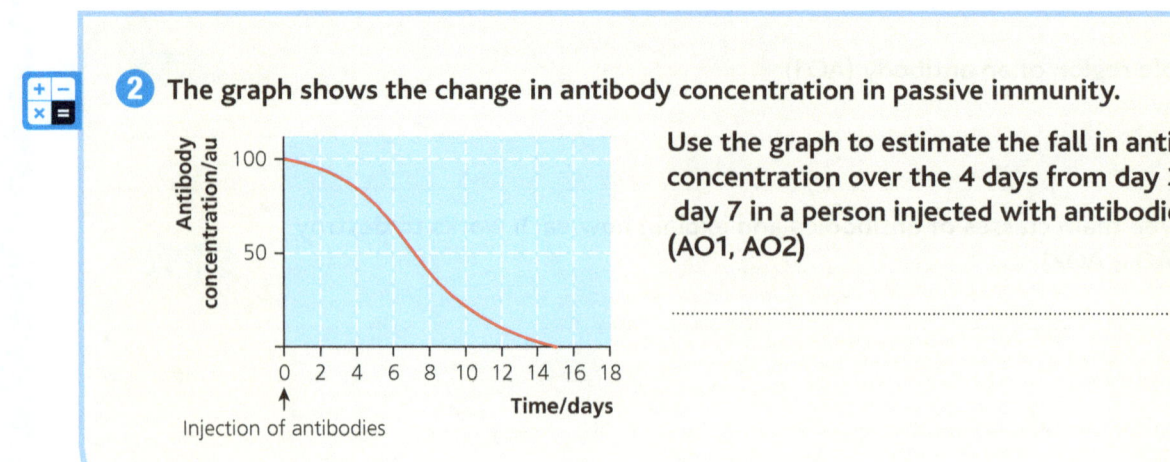

Use the graph to estimate the fall in antibody concentration over the 4 days from day 3 to day 7 in a person injected with antibodies. (AO1, AO2)

2 marks

Autoimmune diseases

An autoimmune disease is one in which the immune system attacks the body's own healthy cells and tissues. The B and T lymphocytes usually respond to antigens on harmful organisms such as bacteria or viruses. However, in an autoimmune disease, the lymphocytes do not distinguish between these 'foreign' antigens and the body's own. As a result, they release antibodies that attack the body's own tissues.

1. Approximately 5% of the population suffers from autoimmune diseases.

    a. What is an autoimmune disease? (AO1, AO2) 1 mark

    b. Suggest what may have gone wrong in someone with this type of disease. (AO1, AO2) 2 marks

Immunisation and vaccination

A vaccination programme is used to help control the spread of infectious diseases and as a means of maintaining the health of the population. Vaccination is given to all individuals, as often as required, from birth to adulthood.

1. What is the difference between vaccination and immunisation? (AO1) 2 marks

2. In order to reduce the impact of further swine flu outbreaks after the 2009 outbreak, a programme of immunisation of certain groups of people was set up. This included young children and the elderly.

a   Suggest two other groups in the population that should be offered immunisation against swine flu. (AO1, AO2) **1 mark**

b   Explain why these individuals are given the vaccine but not other members of the population. (AO1, AO2) **2 marks**

3 Describe the difference between vaccination programmes designed to give ring immunity and those for herd immunity. Suggest the usefulness of each type of immunity in the control of the spread of diseases. (AO1, AO2) **4 marks**

4 Complete the table to show some common problems with vaccines. (AO1, AO2) **4 marks**

Organism	Problem for vaccine development
Plasmodium	
Viruses	
Mutated viruses	
New diseases	

Possible sources of medicines

Among the huge diversity of plants and microorganisms in the natural world, there may be organisms that produce chemicals that are beneficial in fighting a wide range of pathogens or against diseases such as cancer. Every organism that is allowed to become extinct could potentially hold the cure for a major disease. It is therefore vital that we try to maintain biodiversity and conserve as many species as possible.

1 Identify *six* groups of new medicines that are currently being developed. (AO1, AO2) **3 marks**

Antibiotics

Antibiotics were developed from the compounds produced by certain fungi to prevent bacterial attack on the fungal mycelium (the system of branching hyphae) and from certain actinobacteria. Nowadays, antibiotics are usually produced artificially or synthesised by chemical processes using microbes.

1. Suggest why antibiotics are prescribed only to treat bacterial infections. (AO2) **2 marks**

2. There are concerns regarding antibiotic use and development in recent times. Describe the main issues that have raised these concerns. (AO2) **2 marks**

Exam-style questions

1. An investigation into the benefits of active and passive immunisation was carried out on the bacterium *Clostridium tetani*, which causes tetanus. Two groups of patients were used in the trial. Members of group 1 were vaccinated with dead *Clostridium* bacteria and those in group 2 were given antibodies against tetanus derived from another animal.

   The levels of antibodies were monitored in both groups for 80 days. Group 1 showed a high level of antibodies (24 antibody concentration (au)) within 10 days and the concentration then fell steadily to approximately 3 au at 80 days. Group 2 began the study with an antibody concentration of 23 au immediately after the antibodies were introduced, but the level quickly dropped to 2 au at 15 days, where it levelled off before reaching 0 at 20 days.

   What conclusions can be drawn from this experiment about the following?

   a The short-term benefit of both methods. **3 marks**

   b The long-term benefit of both methods. **3 marks**

2. **Complete the table to show some features of non-specific defence mechanisms.** `12 marks`

Type of non-specific defence	Action
Blood-clotting	Rapid response; platelets release a compound; chain reaction or cascade; large number of plasma proteins; fibrin produced; traps blood cells; clot forms
	………………………………… secreted; site of infection inflamed by ………………………………… promoting inflammation; increase in blood flow due to vasodilation; leaky capillaries allow fluid to leave the capillary for tissues; allow plasma proteins to leave; secretion of ………………………………… that stimulate and signal to whole body
	………………………………… at the site divide and begin repair; ………………………………… secrete growth factors; new blood vessels; ………………………………… produced; formation of new tissue; contractile cells contract wound and close it up; destruction of unwanted cells
	………………………………… spread rapidly to the site and destroy pathogens, but then die and form pus; ………………………………… enter tissues and form macrophages; these are important in the specific response; ………………………………… cells have a large surface area to interact and trap pathogens; take them to lymph nodes

3. TB is caused by the bacteria *Mycobacterium tuberculosis*.

   a Describe how *M. tuberculosis* is transmitted in a human population. `2 marks`

   ....

   ....

   b The graph shows the number of notified cases of TB between 1913 and 2006 in England and Wales. Use the graph to estimate the difference in notified cases of TB between 1920 and 2005. `3 marks`

   ....

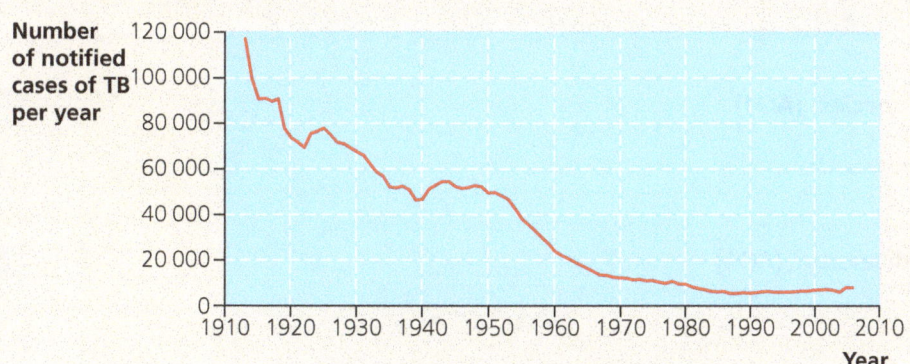

c Describe the changes in TB cases shown in the graph before vaccination and after vaccination was introduced in 1950. `4 marks`

....
....
....

4 The diagram shows an antibody, a molecule that relies on a specific shape to bind to a particular compound.

a Draw a ring around the parts of the molecule that have a complementary shape to the shape of the antigen. `1 mark`

b What are the left and right sides of the structure made of and what is the bond between them? `2 marks`

....
....

5 Vaccinations are effective in preventing the spread of a number of diseases. Why are they an example of active immunity? `2 marks`

....
....

Topic 5 Biodiversity

Levels of biodiversity

Biodiversity is the variety of life. It includes all the different plant, animal, fungus and microorganism species in the world, the genes they contain and the ecosystems of which they form a part. All living things contribute to biodiversity, giving a rich variation in terms of the number of species and the genetic variation of those species. Biodiversity can be considered at three levels: the range of habitats within an ecosystem, the range of species within a habitat and the genetic variation within a species.

1 Define the term *species*. (AO1) `1 mark`

....
....

2 What is species diversity? (AO1) `1 mark`

....

3 Distinguish between species richness and species evenness. (AO1) **1 mark**

....
....
....

4 What determines genetic diversity in different breeds of animals such as the domestic dog or cat? (AO1) **1 mark**

....
....

5 What is the gene pool? (AO1) **1 mark**

....

6 One method of identifying differences in genetic diversity is the study of allozymes, which are enzymes that function slightly differently and affect the phenotype of an individual. These can be identified using electrophoresis. Explain how electrophoresis may be used to study genetic diversity. (AO3) **3 marks**

....
....
....
....
....

7 Complete the table to show some ways of assessing genetic diversity in a population. (AO1, AO2) **3 marks**

Assessment	What this means	Method
Proportion of gene loci with two or more alleles		Determine the genotypes in a sample of the population and calculate the percentage of alleles
Proportion of heterozygotes in population		Cross-breeding or interbreeding (not inbreeding)
The number of different alleles for a gene		Determine differences in protein structure, e.g. blood proteins in dogs show the number of alleles varies between 2 and 11

8 In a study of the horseshoe worm, nine of the 39 gene loci were polymorphic.

    a What is the *P* value for the worm population? (AO3) **1 mark**

b What confidence level would be applied to this calculation? (AO3) `1 mark`

9 What is meant by the term *habitat diversity*? (AO1) `1 mark`

Sampling

Most habitats are fairly large and contain large numbers of plants and animals. It is therefore impossible to count the number of individuals in each species or all living things in one place. Sampling techniques are therefore important. These involve studying small parts of the habitat in detail and then using this to estimate the size of the entire population.

1 Which sampling technique is best employed when investigating the plant species in a grassland habitat? (AO1, AO3) `1 mark`

2 Suggest in what type of investigation a non-random line or belt transect may be the best method of sampling. (AO1, AO3) `1 mark`

3 Sampling animals requires other techniques because they move and may only appear at night. Suggest two other sampling techniques that may be more useful in sampling animal populations. (AO1, AO2) `2 marks`

Simpson's Index of Diversity

Simpson's Index of Diversity measures both species richness and species evenness. To calculate diversity, use the following formula:

$D = 1 - (\Sigma(n/N)^2)$

where Σ = 'the sum of', n = the total number of individuals in a particular species and N = the total number of individuals in all species.

1 A student carried out an investigation to collect data and calculate the Simpson's Index of Diversity in an area of grassland using randomly placed quadrats. Five different species were identified, excluding grass and moss.

    a Carry out the calculations to complete the table. (AO2, AO3) `3 marks`

Species	Total number of each species (n)	n/N	(n/N)²
Dandelion	27		
Clover	7		
Trefoil	1		
Plantain	1		
Groundsel	3		
Total	N =		Σ =

b Calculate Simpson's Index of Diversity (*D*), showing your working. (AO2, AO3) `1 mark`

c What conclusion can be drawn from the *D* value obtained in part b? (AO3) `2 marks`

....

....

....

....

Factors affecting biodiversity

Biodiversity is affected by the daily activities of living things, especially humans. Human population growth is the biggest threat of all.

1 Complete the table to show the factors, causes and effects for the changes in biodiversity. (AO1, AO2) `5 marks`

Factor affecting biodiversity	Causes	Effects
Deforestation	Clearing for:	Loss of .... and loss of resources to rebuild number of species; loss of ....
Destruction of coral reefs	Farming, ...	
Destruction of the sea bed	Dragnets scrape the sea bed	
	Removal of timber for industry or furniture; overfishing so fish stock falls	Reduction in the number of species and loss of habitat
	Removal of wild animals as 'bush meat'	Species at risk of extinction

Factor affecting biodiversity	Causes	Effects
Agriculture	.... of one type of crop or livestock	Reduction in the number of a particular species
Pollution from agriculture	Waste products; use of	Loss of .... as some species thrive with additional fertilisers and others are destroyed; runoff pollutes streams and pond water
	Modification of weather patterns such as drought and floods affects species distribution	Distribution of species changes; lack of food for some species causes decline

Maintaining biodiversity

Conservation is the protection and maintenance of ecosystems and the species within them. Ideally, conservation is in the natural habitat of the species (*in situ*), although this is not always possible. International organisations have designated some areas for conservation, such as the biosphere programmes in the Cors Dyfi nature reserve in Wales. World heritage sites are those identified as culturally or physically important and include St Kilda off northwest Scotland. There are also National Parks, national reserves and marine conservation zones.

1 Explain the biological and economic reasons for maintaining biodiversity. (AO2) `4 marks`

2 What do the protected sites mentioned in the text above have in common? (AO2) `3 marks`

3 *In situ* conservation may not be possible where the original habitat is too small or fragmented, or the species number is too low. Describe two other methods of conservation that may be used to offer protection. (AO2) `2 marks`

Exam-style questions

1 A study of plants in the Area of Outstanding Natural Beauty on the Sussex Downs was carried out to determine species richness.

   a What is species richness?   `1 mark`

   b Describe how a sampling procedure could be carried out to ensure that a representative sample is taken.   `3 marks`

   c What is the importance of species evenness in determining biodiversity in a specific habitat?   `3 marks`

   d In a comparison between two sites, student A obtained an Index of Diversity of 0.68, whereas student B (working on another site) obtained an Index of Diversity of 0.54. Comment on the significance of these two values and the sites at which they were obtained.   `1 mark`

2 The Royal Botanic Gardens at Kew is an important provider of plant conservation. One plant conserved there is the thermal water lily, *Nymphaea thermarum*, which grows naturally in hot springs in central Africa.

   a In what genus does the thermal water lily belong?   `1 mark`

   b Explain why it may be necessary to conserve a species such as the thermal water lily outside its natural habitat (*ex situ*).   `3 marks`

   c The Millennium Seed Bank at Wakehurst Place, Sussex is another branch of Kew. Give two advantages of conserving plant species as seeds and not as adult plants.   `2 marks`

d In order to measure biodiversity in a particular habitat it is usually necessary to sample a proportion of the habitat. Give an outline of an unbiased sampling method that could be used to measure biodiversity in a downland habitat such as the South Downs.

4 marks

....

....

....

3 The graph shows catches of Atlantic cod, *Gadus morhua*, between 1950 and 2012.

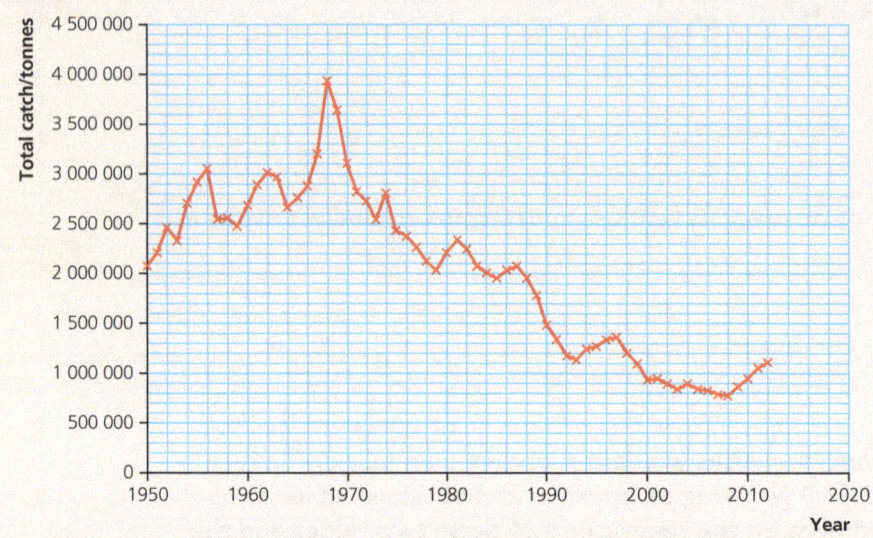

a Use the graph to calculate the rate of decline in catches between 1968 and 1970 and compare this with the rate of decline between 1997 and 1998. (Hint: use the gradient of the line.)

2 marks

b Explain what the decline in fishing catch indicates about the cod fish stock and describe the conclusion that can be drawn from this information.

2 marks

....

....

c From 2008 to 2012, the catch increased. Make *three* suggestions to explain this increase.

3 marks

....

....

....

4. African elephant populations are at risk of extinction. Testing the DNA found in ivory allows the geographic location of the elephant population from which the ivory was taken to be determined.

   a. Other than ivory, give *two* other possible sources of elephant DNA. **2 marks**

   b. Using ivory DNA, it is possible to trace the specific living populations of elephant. What conclusion can be drawn about the population dynamics of the African elephant? **2 marks**

   c. African elephants were registered on the CITES list for endangered animals in 1990. What does CITES stand for? **1 mark**

   d. Give *three* potential benefits of registration with CITES for conservation of the African elephant. **3 marks**

5. Although inorganic fertilisers are considered to be non-toxic to living organisms, excessive use of these fertilisers can lead to a reduction in biodiversity of the farmland and surrounding areas.

   a. Suggest how the excessive use of inorganic fertilisers on farmland can cause a reduction in biodiversity. **2 marks**

   b. Explain why a reduction in biodiversity may create problems for agriculture in the future. **3 marks**

   c. How might the use of the fertilisers on farmland affect the surrounding land? **2 marks**

Topic 6 Classification and evolution

The biological classification of species

Classification is a way of grouping organisms according to shared characteristics. A natural classification involves a variety of different methods including how organisms have evolved, whilst an artificial classification is one relying solely on identifying features that may not reflect how the organisms have evolved. Taxonomy is the study of classification and grouping of organisms within a ranked or hierarchical system.

1 List the different ranks of groups, or taxons, in order. (AO1) — 4 marks

2 To which class do humans, whales, cattle and dogs all belong? (AO1) — 1 mark

3 *Giraffa jumae* is an extinct relative of the modern giraffe and may have been close to its ancestor. Give the hierarchical classification of *Giraffa jumae*. (AO1, AO2) — 1 mark

The binomial system

The binomial system is the way in which two Latin words are used to name each species. The first name is the name of the genus to which the species belongs and the second name is the specific or species name. Both of the Latin names should always be written in italics or underlined. The genus name should be written with an upper case first letter, whereas the species should be in lower case.

1 The Galapagos mockingbird (*Mimus parvulus*) has many similarities and some differences with the Española mockingbird (*Mimus macdonaldi*).

   a The Galapagos mockingbird has two names. What is the name of its genus? (AO1) — 1 mark

   b Give the Latin name for the Española mockingbird and explain why one name has a capital letter and the other does not. (AO1) — 2 marks

The five kingdoms

Carl Linnaeus (1707–78) devised the five kingdoms of classification as the first true system of classification. This uses the organisms' observable features and groups them according to the number of similarities present.

1. Name the five kingdoms and explain the system. (AO1, AO2) `1 mark`

   ....

   ....

   ....

2. Complete the table to show the characteristics of the five kingdoms. (AO1, AO2) `8 marks`

Features	Kingdoms				
	Prokaryotae	Protoctista	Fungi	Plantae	Animalia
Body type	Mostly unicellular	Unicellular and multicellular	Mycelium of hyphae (except yeasts)		
Nucleus	Absent				
Cell walls	Wall of peptidoglycan	Wall present in some			
Cell vacuoles	Only in a few	Some (e.g. algae): large, permanent vacuole. Protozoans: small, temporary vacuole			
Organelles			Present	Present	Present
Motility	Some have flagella				
Nervous coordination	Absent				
Examples (give two of each group)					

3. For each of the five kingdoms, give the main characteristics that distinguish them from the other groups. (AO1, AO2)

   a **Prokaryotae** `1 mark`

   ....

   ....

   ....

   b **Protoctista** `1 mark`

   ....

   ....

   ....

   c **Fungi** `1 mark`

   ....

   ....

   ....

d   Plantae 1 mark

e   Animalia 1 mark

New classification systems

Until some 50 years ago, classification systems were based on the anatomy, physiology, morphology, cell structure and behaviour of an organism. However, the development of molecular biology and macromolecules has provided the ability to use molecular phylogeny. The three domains of life classification of Bacteria, Archaea and Eukaryota are now widely accepted and the domains are a taxonomic level above the five kingdoms.

1 The classification of organisms into three domains is quite recent. Outline the main characteristics and differences between a domain-based and a kingdom-based classification. (AO1, AO2) 4 marks

2 Write a short synopsis of each of the main types of molecular phylogeny used today. (AO1, AO2)

a   antibodies 2 marks

b   protein sequencing 2 marks

c   DNA hybridisation 2 marks

d DNA sequencing *(2 marks)*

Classification and phylogeny

Natural classification groups living things according to their similarities — the more features that are shared between two organisms, the more closely related they are. Phylogeny is the evolutionary history or the evolutionary relationships between organisms and groups of organisms. Therefore, natural classification reveals the phylogeny.

1 Phylogeny is only reflected in natural classification systems. Define the word *phylogeny*. (AO1) *(2 marks)*

Evidence for the theory of evolution

The theory of evolution supports the idea that organisms change over a long period of time as a result of selection pressures caused by natural selection acting on variations in the population. Darwin called it 'descent with modification'.

1 Darwin proposed that those organisms best suited or adapted to their environment were most likely to survive and pass on their characteristics to future generations. What are the main principles suggested for the ability to survive by natural selection? (AO1, AO2) *(3 marks)*

2 Write a summary of these lines of evidence for evolution. (AO1, AO2)

  a morphology *(2 marks)*

  b anatomy *(2 marks)*

    c   fossils   2 marks

    ....
    ....
    ....

    d   biochemistry   2 marks

    ....
    ....
    ....

    e   classification   2 marks

    ....
    ....
    ....

Variation

Variation refers to the differences that exist between individuals. Intraspecific variation occurs between members of the same species, whereas interspecific variation occurs between members of different species.

Continuous variation is seen where there are no distinct groups or categories. There is a full range between two extremes. Examples of continuous variation include height and body weight. Discontinuous variation is where there are distinct groups or categories and no in-between types. This type of variation is usually caused by one gene (or possibly a small number of genes). Examples include gender and possession of resistance or immunity.

1 Compare genetic variation with phenotypic variation. (AO1, AO2)   2 marks

....
....
....

2 What is the difference between interspecific and intraspecific variation? (AO1, AO2)   1 mark

....
....

3 Complete the table to compare continuous and discontinuous variation. (AO1, AO2)   2 marks

Type of variation	Continuous	Discontinuous
Discrete categories		
Intermediates		
Continuous with no categories		

Type of variation	Continuous	Discontinuous
Type of data		
Genetic variation		
Environmental effect		
Examples		

4 An experiment was conducted into the number of spines on holly leaves. Fifty leaves were collected and measured from one holly tree. The experiment was repeated three times and then repeated three times on a different species of holly tree. The results are given in the table. *2 marks*

Holly leaf type	Replicates of spine numbers			Mean number of spines
Ilex aquifolium	12	14	9	
Ilex verticillata	3	2	4	

a Complete the table by calculating the mean number of spines per holly leaf using the data provided. (AO2, AO3) *1 mark*

b What errors have been made in recording this data? (AO2, AO3) *2 marks*

c The student concluded that *Ilex aquifolium* has a greater number of spines than *Ilex verticillata* on its leaf margin. Discuss the validity of this conclusion given the data on which it is based. (AO2, AO3) *2 marks*

The implications of evolution for humans

Modern examples of evolution include pesticide resistance in insects and drug resistance in microorganisms.

1 State the similarities and differences between antibiotic drug resistance and pesticide resistance. (AO1, AO2) *3 marks*

Exam-style questions

1 Chimpanzee DNA is 98.4% identical with human DNA.

a Suggest how DNA analysis may be useful to taxonomists.

2 marks

b Give *two* types of evidence other than biochemical evidence that may be used by taxonomists when classifying organisms.

2 marks

2 a Cytochrome C is a protein found in living organisms. The structure of this protein varies between different organisms. However, closely related organisms have similar cytochrome C protein structure. Suggest where cytochrome C may be most likely to be found in most living organisms.

1 mark

b Placental mammals and marsupial mammals share a common ancestor and have very similar cytochrome C protein structures. From your knowledge of classification, in which two of the following groups of egg-laying vertebrates would you expect to find the most similar cytochrome C protein structures?

2 marks

cartilaginous fish   bony fish   amphibians   reptiles   birds   egg-laying mammals